WHEELS & THINGS

Memories of Yesteryear

By the same author:

"Erk's Eye View" *A story of National Service in the Royal Air Force during the 1950's. Brewin Books 1995.*

By Eric W. Russell

BREWIN BOOKS

First Published in October 1996 by
BREWIN BOOKS, Studley, Warwickshire.

ISBN 1 - 85858 - 093 - 5

British Library Cataloguing in Publication Data
A Catalogue record for this book is available from the British Library

Typeset in Garamond and Printed in Great Britain by
Heron Press, Kings Norton
Birmingham B38 9TS

This book is dedicated to my late parents, Lydia and Eric.
Despite the years of my childhood and informative years (1930's & 40's) being a difficult period in our history, they succeeded in leaving me a legacy of happy memories.

Author with Parents at Bossington, 1938.

PREFACE

When I started to plan this book, I set myself the task of creating a work of pure nostalgia, together with an element of informative entertainment. Over the past few years, many excellent books have been produced that have dealt in great detail with things of which I only make brief mention. My main objective was to bring together a small collection of pictures, designed to reflect some of the familiar sights of my childhood. A childhood that started in the good years of the mid 1930's, through the years of the Birmingham Blitz of the early 1940's, to becoming a teenager shortly after the end of the second world war.

I realised that unless I set myself a specific theme, things would get out of hand. I quickly decided it must be a theme that would allow me reasonable degree of content flexibility, (artists license). I then hit on a general theme of "Wheels", this seemed perfect as I became an enthusiastic cyclist during my youth and eventually became an apprentice in the motor industry; more wheels. I only had one more problem before I actually got down to serious writing, that was to choose a title that would allow me the flexibility to include material that had no direct relationship to wheels. After much thought it suddenly came to me, the answer had in fact been staring me in the face; the title would be . . . . .

WHEELS AND THINGS

THE AUTHOR IN 1943

The "Wheels" in this picture are a Hornby '0' gauge locomotive.
It was a GWR castle class, Caerphilly Castle. It was very much a toy in 1943, but would now command a high price as a collectors item.
Oh how times and values have changed.

The Wall's "Stop Me and Buy One" ice cream vendors were a familiar sight in the 1930's. The company had a nationwide network of depots. The 'Snofrute' was the forerunner of the now familiar ice lolly. The Snofrute was contained in a triangular shaped paper tube, similar to today's Toblerone wrapper. The vendor always had a small knife to hand so that he could cut a Snofrute in half; he would sell the two halves at 1/2d each. Oh the happy days of youth of yesteryear.

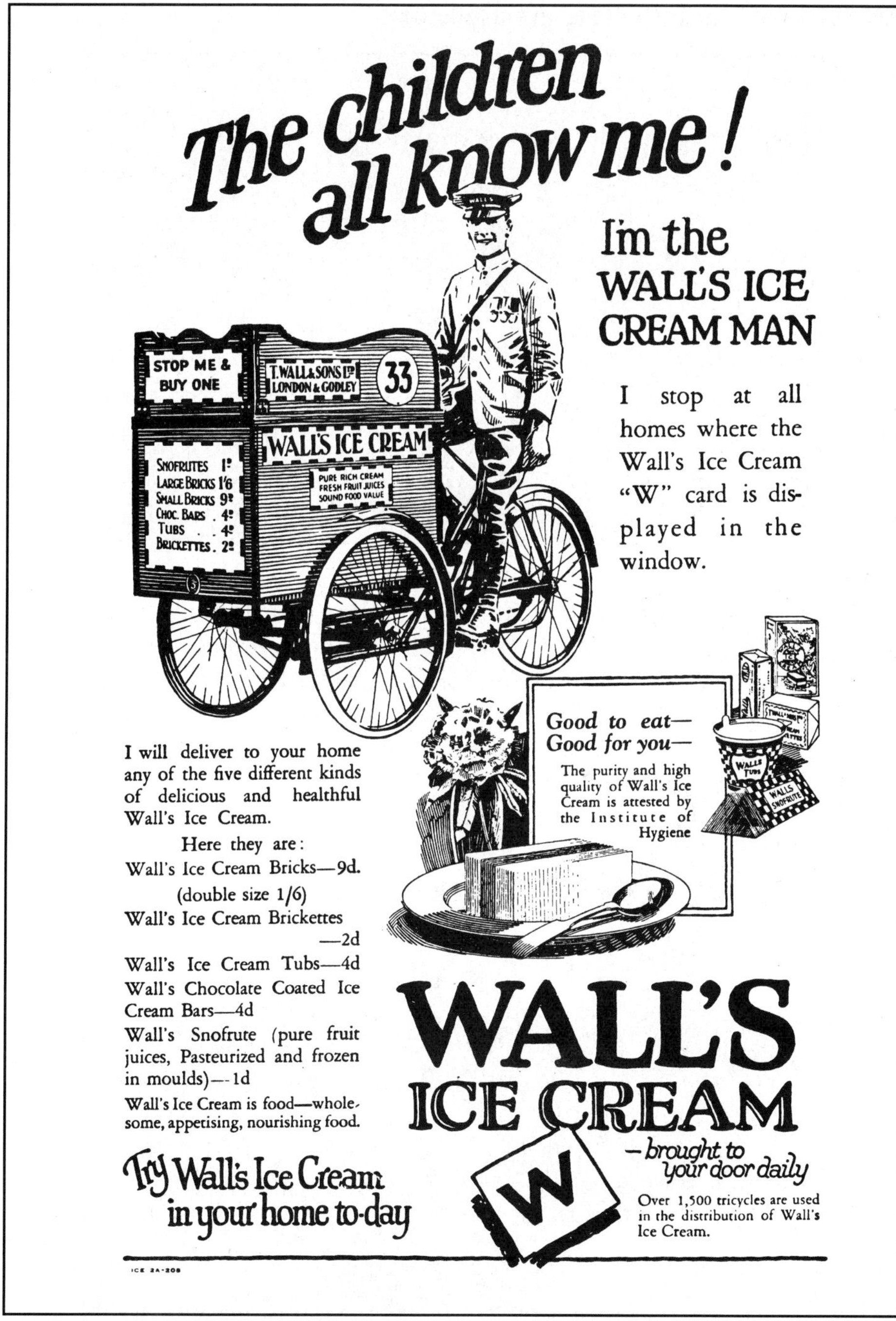
The children all know me!
I'm the WALL'S ICE CREAM MAN
I stop at all homes where the Wall's Ice Cream "W" card is displayed in the window.
STOP ME & BUY ONE
T. WALL & SONS LTD LONDON & GODLEY
33
WALL'S ICE CREAM
PURE RICH CREAM FRESH FRUIT JUICES SOUND FOOD VALUE
SNOFRUTES 1d
LARGE BRICKS 1/6
SMALL BRICKS 9d
CHOC. BARS . 4d
TUBS . . 4d
BRICKETTES . 2d
Good to eat— Good for you—
The purity and high quality of Wall's Ice Cream is attested by the Institute of Hygiene
I will deliver to your home any of the five different kinds of delicious and healthful Wall's Ice Cream.
Here they are:
Wall's Ice Cream Bricks—9d. (double size 1/6)
Wall's Ice Cream Brickettes —2d
Wall's Ice Cream Tubs—4d
Wall's Chocolate Coated Ice Cream Bars—4d
Wall's Snofrute (pure fruit juices, Pasteurized and frozen in moulds)—1d
Wall's Ice Cream is food—wholesome, appetising, nourishing food.
Try Wall's Ice Cream in your home to-day
WALL'S ICE CREAM
—brought to your door daily
W
Over 1,500 tricycles are used in the distribution of Wall's Ice Cream.

Cyclo cross as it used to be. The annual Rugby November Handicap, about 1949. The author, together with many of his old friends of the Warwickshire Road Club are in this picture.

Scott's cycle shop, Coventry Road, Birmingham. This business was on the corner of Wordsworth Road. The business and premises were bought from J. Woodgate shortly after the tornado of 1931. Scott's stayed on the site until 1970. The business then moved to Stratford Road. The present owner (1995) is a grandson of the man who purchased the business in 1931.

The author (left) with club mate, Bob Dunn at the Birmingham Charity Sports 1951. This cycling and athletics meeting was held annually at the BSA Sports Ground in Golden Hillock Road.

Golden Jubilee Rally of the Birmingham & Midland District Association of the Cyclists Touring Club, (CTC). Meriden 12th June 1949.

Russell's ironmongers was my grandfathers business. It was located on the Stratford Road at Shirley. The billhead of the 1930's gives the address of 100 and the district as B'ham 11. Properties on Stratford Road were later renumbered and it became 110. The picture was taken about 1947 and shows a well stocked window plus a traditional pavement display, with no chance of any stock being stolen. Oh how times change.

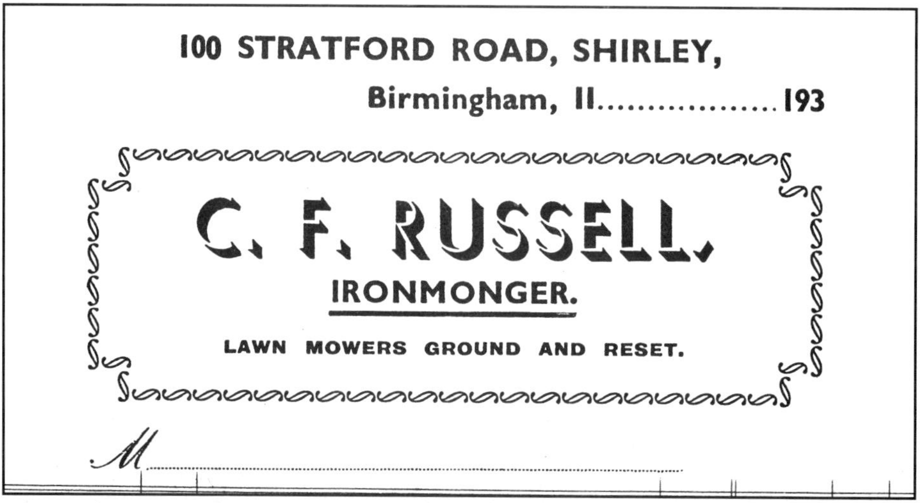

100 STRATFORD ROAD, SHIRLEY,

Birmingham, 11.................193

C. F. RUSSELL,

IRONMONGER.

LAWN MOWERS GROUND AND RESET.

M...

Manor Farm Shirley was the home of the Benton family. The farm was facing my grandfathers shop on a site between Olton Road and what is now Jacey road.Joseph Benton ran his coal business from the farm for several years before moving to Tilehouse farm. The child in the upper picture being held by his sister, is Francis Benton, who now runs the Benton haulage business. The lower picture shows Joseph Benton standing by his coal wagon outside Robin Hood Cemetery. His companion is Arthur Key.

Mrs. Badger taught the piano at her home which was next to the premises of Morris and Jacombs, builders. I had lessons from her in the early 1940's and admit to being one of her failures. I never mastered a musical instrument, but did develop an ear for listening to good music which has given tremendous pleasure.

Home Leigh school was located in a large wooden building in the grounds of St. Benedicts church. I attended the school for a short time in the early part of the war as my own school, Starbank Road, was closed for a short period due to the evacuation of some pupils and staff. Miss Westwood was assisted at Home Leigh school by Miss Irene Badger, the daughter of Madam Grace Badger.

Bottom picture shows the site of Home Leigh school in 1989.

HOME LEIGH SCHOOL

S. Benedict's Parish Room, Small Heath.

Principal—Miss Westwood.

Three Terms per Year—Christmas, Easter, and September.
Modern Methods, English, French, Music, Physical Culture,
Kindergarten, Preparation for School Examination.
Recent Successes. Terms Moderate.

Madam Grace Badger, L.T.C.M. (Senior) Honours T.C.L.

Teacher of the Pianoforte & Theory of Music

Tamar House, 73, Hobmoor Rd., Small Heath.

PUPILS EXAMINATION SUCCESSES FOR 1933—

April—Irene Badger - Intermediate Honours, London Col.
July—Irene Badger - Junior Honours Medalist, Trinity Col.
Doris Macdonald - Elementary First Class, London Col.
Gene McGrah - - - - - - - Primary First Class
Joyce Heath Primary Pass. Jessie Anson Primary Pass
Constance Howes - - - - - - - - Primary Pass
Irene Evans - - - - - Primary First Class Theory
Ronald Andrews - - - - - Primary Pass Theory
Doris Macdonald - - - - - Primary Pass Theory
December—Amy Bourton - - - - - - Senior Pass
Irene Badger - - Advanced Intermediate, First Class
Doris Macdonald - - - - Intermediate First Class
Irene Badger - - - - - Primary First Class Theory
Ronald Andrews - - - - Elementary Pass Theory

In January of 1944 my mother took me to see the pantomime, Robinson Crusoe at the Birmingham Theatre Royal. This was an old, much loved theatre that was situated at the top of New Street. I remember particularly the performance of Nat Jackley, The India Rubber Man. I was to see him many times in later years, both on stage and television. I readily admit that I always found his unsophisticated slapstick comedy amusing. During that far off panto of 1944, Nat jackley came into the auditorium and chatted to the audience. My mother asked him to autograph her programme so that she could send it to my father, who at that time was on active service in the middle East. On hearing her request he took the programme backstage and obtained autographs and signed photos of several members of the company. Truly a gentleman of the theatre. Below are two of the signed photos from 1944.

On the following two pages are pictures of that battered and much travelled programme of yesteryear. It came back to Birmingham in 1945 having journeyed through Tunisia, Italy, Palestine and Egypt. On its front cover is the singer and entertainer of yesteryear, G H Elliott, 'The Chocolate Coloured Coon'.

MURRAY & MOONEY.

PRIESTLEY'S
for
PIANOS
and
Radio of course!
In fact—your every Musical need
W. H. PRIESTLEY
AND SONS LTD.
112, NEW STREET, BIRMINGHAM
Selected hops, finest barley, yeast and sugar make this fine flavoured beer
Mitchells & Butlers
"Good Honest Beer"
THERE'S AN "M&B" HOUSE QUITE NEAR TO YOU
THIS SEAL IS UNBROKEN
"Thank goodness
I brought some T.C.P!"
A nasty knock — but she knows that T.C.P. will keep it safe from dirt and germs. How valuable T.C.P. has become, now that all of us are doing war-time jobs of one kind or another.
T.C.P. can be used—any time, anywhere—for almost every common ailment and emergency. For though it destroys infecting germs instantly, T.C.P. is much more than 'just an antiseptic' in the usual sense. It relieves pain, it reduces inflammation, it actually assists clean, scarless healing. And it is so completely SAFE that you can use T.C.P. on any part of the body — including the eyes, nose and mouth, and even take it internally, to counter infection in the system itself.
No wonder men and women who want to keep 'on the job' always have a bottle handy. T.C.P., with full instructions, is sold by all Chemists in 1/3d. and 3/- bottles. Purchase tax extra.
T.C.P
'ERYDAY EMERGENCIES
Co. Ltd., Advertising Contractors, London
dington Ltd., Leeds: and London
BIRMINGHAM
Theatre Royal
2d
PLEASE SEE THAT
Programm

BIRMINGHAM THEATRE ROYAL

Telephone .... MIDland 4355-6-7
Stage Door .... MIDland 4053

Chairman .... R. H. GILLESPIE
Managing Director .... GEORGE BLACK
Direction .... GEORGE BLACK
Manager .... C. F. PRICE
Musical Director .... JOHN STONE
Press Representative .... LAWSON E. TROUT, F.Inst.P.
(MOSS' EMPIRES LTD.)

In accordance with the requirements of the Licensing Justices:
(a) The public may leave at the end of the performance by all exits and entrances other than those used as queue waiting-rooms, and the doors of such exits and entrances shall at the time be open.
(b) All gangways, passages and staircases shall be kept entirely free from chairs or any other obstructions.
(c) Persons shall not be permitted to stand or sit in any of the intersecting gangways. If standing be permitted at the rear of the seating, sufficient space shall be left for persons to pass easily to and fro.
(d) The fireproof curtain shall at all times be maintained in working order and shall be lowered at the beginning of and during the time of every performance.

"Alert" and "All-Clear" Signals will be advised to patrons by means of illuminated signs which are placed at each side of the Stage. Those desiring to leave may do so, but the performance will continue.

PRICES OF ADMISSION

(Including Entertainment Tax)

Boxes 48/- and 28/-, Orchestra Stalls 8/3, Royal Stalls 7/-, Stalls 5/9, Pit Stalls 4/6, Dress Circle 7/6, Circle 6/3, Upper Circle 4/-, Balcony 1/6

REDUCED PRICES FOR CHILDREN at all Matinees from Monday, January 17th, 1944, will be as follows:

Orchestra Stalls 5/9, Royal Stalls 5/-, Stalls 4/6, Pit Stalls 3/6, Dress Circle 5/9, Circle 4/6, Upper Circle 3/6, Balcony no reduction

PARTY BOOKINGS:

As from January 24th, 1944, Parties of 30 or more will be admitted to Orchestra Stalls, Stalls, Dress Circle and Circle (exclusive) at all performances except Saturdays.

Commencing Christmas Eve, December 24th (One Performance) at 6.0 p.m. No performance on Christmas Day. Evenings at 6.0 p.m. Matinees at 2.0 p.m. The first Matinee will be on Monday, December 27th. Afterwards Twice Daily until Saturday, January 15th, 1944. From that date Matinees on Monday, Wednesday, Thursday and Saturday until further notice.

PROGRAMME

TOM ARNOLD

(For Julian Wylie Productions Ltd.)
presents

The Royal Pantomime

Robinson Crusoe

Book by Dan Leno Junr., Robert Nesbitt and Tom Arnold

Produced by **Robert Nesbitt**

CHARACTERS

Robinson Crusoe ...... **Betty Frankiss**
Polly Perkins .... **Sybil Summers**
Billy Crusoe .... **Henry Ford**
Will Atkins .... **Nat Jackley**
Mrs. Crusoe .... **Eddie Henderson**
1st and 2nd Mates **Murray and Mooney**
Sambo .... **G. H. Elliott**
Bosun .... **Richard Beamish**
Cannibal King .... **George Fairweather**
Man Friday .... **Harry Eades**

Principal Dancer: **Joy Caville**

The Three Leopard Dancers

The Loretta Dancers

Chorus and Ballet

SYNOPSIS OF SCENES:

PART I

SCENE 1 .... THE PORT OF HULL
SCENE 2 .... ON BOARD—THE HOLD
SCENE 3 .... ANOTHER PART OF THE SHIP
SCENE 4 .... THE MAIN DECK

TABLEAU: THE OPEN SEA

— INTERVAL —

Orchestra under the Direction of

Mark Chesnay

PART II

SCENE 5 .... THE ISLAND
SCENE 6 .... THE SEASHORE
SCENE 7 .... THE STOCKADE
SCENE 8 .. THE VISION—RULERS OF THE SEA
SCENE 9 .... BACK IN HULL
SCENE 10 .... CRUSOE'S NEW SHIP

Costumes designed by Doris Zinkeisen and executed by L. & H. Nathan Ltd., Alec Shanks and Betty S. Roberts. Electrical Effects by Strand Electric Co. Ltd.

Manager and Stage Director .... **W. E. Barratt**
Wardrobe Mistress .... **Mrs. Baines**
(For Julian Wylie Productions Ltd.)

Fire Engine & Pump of yesteryear. Pumps of this type became a familiar sight during the blitz era of the early 1940's. They would be towed by a wide assortment of motors, ranging from Austin Seven cars to heavy commercial vehicles.

Wooden bodied refuse wagons were an everyday sight in the 1930's & 40's. Due to the behaviour of unthinking and uncaring householders who deposited hot ashes in their dustbins, the bodies of the vehicles often caught fire when the ashes were tipped in.

The AA and RAC patrol men with their distinct liveried motorcycle combinations were a familiar sight on the roads. That was in the days when motoring was unhurried and a pleasure, it was in fact an adventure at times. The speed and volume of traffic allowed patrol men to salute members vehicles. Failure to get a salute often indicated that the member should look out for a hazard, such as an accident or indeed a police speed trap.

The premises of Vale Onslow, a long established Birmingham based motorcycle business. The brand names above the shop would suggest that the picture was taken during the heyday of the British motorcycle industry. Not a Japanese name to be seen.

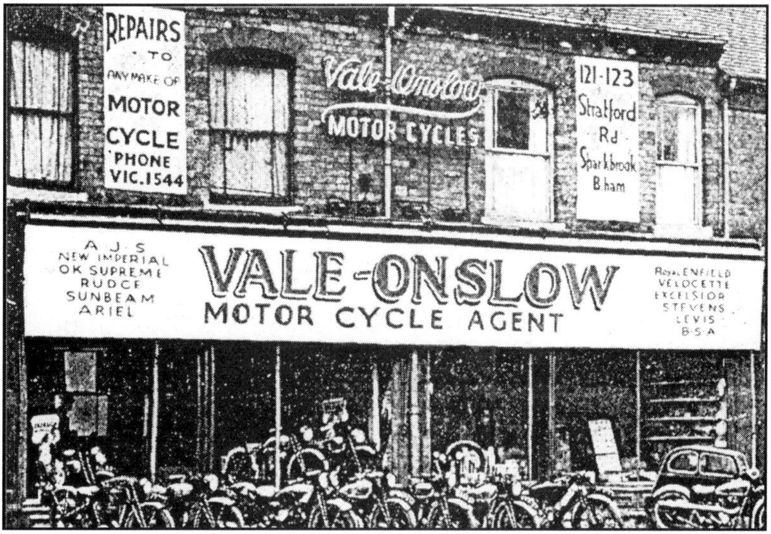

The motorcycle combination was the property of the late Len Udall. Believed to be a Velocette MSS 500cc of about 1937. This would be logical as the Udall family had a long association with the Velocette company at its Hall Green, Birmingham factory.

Advertising page from St Benedicts church magazine of August 1930. The only wheels in this picture are on the pram which was priced at £8-8-0. Interesting to see that all the items were post free, with the exception of the bible which was plus 6d; strange!

CHURCH ASSEMBLY NOTES AUGUST, 1930

SELFRIDGES

For All Postal Shopping

You may order any of these Bargains by post or telephone with perfect confidence. Satisfaction is assured or your money will be refunded without quibble. Post Orders are executed by specially trained assistants who are skilled in interpreting customers' requirements.

A REMARKABLE BOOK OFFER

HANDSOME BIBLES *in Coloured Bindings*

A very handsome BIBLE which is exclusive to Selfridge's Book Department. Bound in a very hard-wearing, new material that resembles the finest Morocco leather but will give almost better wear. In 8 colours, Violet, Purple, Maroon, Green, Blue, Scarlet, Brown or Black. Very light in weight, being printed on very thin paper, but the type is **clear and bold.** Full gilt edges, and bookmarker. 8 maps in Colour. A most remarkable offer at the price of **5/-** each

Postage 6d. extra.

Ground Floor, Aisle 8.

Orders 10/- and over, Post Free.

Start them right with an "Adonis" Baby Carriage.

A smart and roomy carriage measuring 33½" by 16" on bed level, generously upholstered with well padded seats, which are loose and easy to clean. The hood and storm aprons are of black leather duck having corner protectors, braid to front of hood and heavy quality black covered joints. Wheels are ball-bearing, tangent spoke, easy riding, well tempered springs, without straps. Body has lined safety strap and wheels fitted mudguards. Colours: Navy/Navy, Navy/Fawn. PRICE each **£8 8s.**

Or delivered after the first of 12 monthly payments of 14/8. Carriage Paid in England and Wales.

Perambulator Dept.—Second Floor.

BEACH AND BATHING WEAR

TOWELLING BATHING CAPES

Wonderful offer for Women. These capes are absolutely indispensable for bathing. They comprise all colours and can be worn with any colour bathing costume. PRICE, each **3/6** Post Free.

Bargain Basement, Aisle 18B.

Wool Stockinette BATHING COSTUMES

Well cut and made to represent the popular TWO-PIECE STYLE, with light top and dark knickers. White belt at waist and available in Royal/Black, Green/Black, Red/Black, Orange/Black, also in Black or Navy. PRICE, each **3/11** Post Free.

Bargain Basement, Aisle 19B.

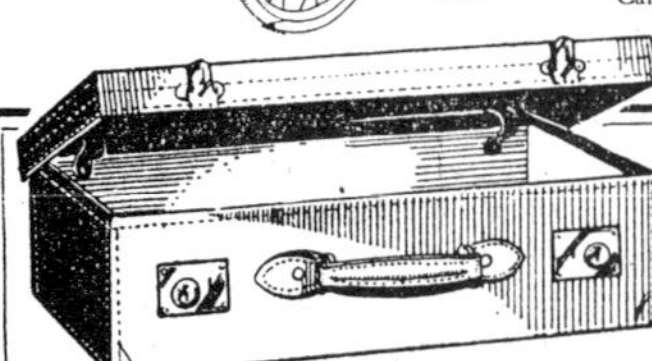

The "Traveller" Suitcase

will endure hard wear, though it is quite light to carry. Made of Cowhide on fibre foundation, steel frame, strongly sewn edges. Each fitted with two strong double action locks, and lined with a cotton material. Size, 24 × 13¼ × 6 inches. Post Free. PRICE, each **25/-**

Trunk Department, First Floor.

Selfridge & Co., Ltd. Oxford St. London, W.1

There was always great excitement when steam (road) rollers came to make up or repair local roads. These two fine specimens were photographed in the 1940's and belonged to the Clee Hill Transport Company. This company was responsible for the make up and maintenance of many miles of roads in the counties of Hereford, Worcester and Shropshire. The company enjoyed a reputation for its outstandingly high standard of work.

A rare 1930's aerial photograph of the Swan Island at Yardley. Centre left is the Swan hotel, vastly different and visually more pleasing than the monstrosity that now graces the site. Across the island from the Swan a trolley bus is visible. The factory style building behind the bus was "Colliers Corner", garage and motor work shops. In the left foreground is Harding's bakery which was in Church Road.

HARDING'S "IDEAL" BREAD

has been awarded 5 Challenge Cups and 50 Gold Medals.

—ALWAYS AT THE TOP FOR QUALITY.—

Ask van to call or send Post Card to—

Royal Steam Bakery, Yardley.

Telephone: ACOCKS GREEN 251.

I was familiar with the Small Heath Gardeners club from an early age, as I was born a few doors away from the club premises in Hobmoor Road. My grandfather, William Maund and our next door neighbour, Sam Richards had been two of the original founder members. The earlier picture shows the club, complete with wooden fence, as I remember it from my early childhood. The later photo shows the premises in 1989. Fence long gone, a concert and dance room above the rear of the premises and a facade to the building which in my opinion is an architectural monstrosity and the epitome of poor taste.

Birds was a very old established Birmingham company with a large factory and distribution complex in Digbeth. The company is now part of a large food group and sadly no longer based in Birmingham.. Their Mercedes delivery van must have been a rarity in the midlands back in 1936, it no doubt turned a few heads as it passed by. I have no details, but I assume that the vehicle would have been withdrawn from service at the outbreak of war in 1939. The Birds advert dates from 1930.

The Lucas Company has its origins in Birmingham. Its original strength came from its King of the road cycle and motor lamps. The company is now a major world wide supplier of components to the motor and aerospace industries.

Lee Longland have long enjoyed a reputation for excellence since its founding in 1902. The business started in premises facing the Hall of Memory in Broad Street. They later moved further along the street to the custom built premises they now occupy. A premises that with its Portland stone facing has more the appearance of a municipal building than a furniture retailer.

Dunlop have produced a wide variety of products over many years but are perhaps best known for their tyres and golf balls. The company premises at Fort Dunlop, dominated the Erdington skyline for many years. Considerable modern development has meant that Fort Dunlop now has to compete with many more buildings for its share of the skyline. I particularly like the tyre advert, (bottom picture) as it depicts what I term "proper racing cars".

A consignment of Cadburys Bourneville manufactured confectionery being delivered to a typical 1930s style store.

No 6 delivery van of Holdens Paints. Pictured in 1935 with Mr. Jack Billington at the wheel.

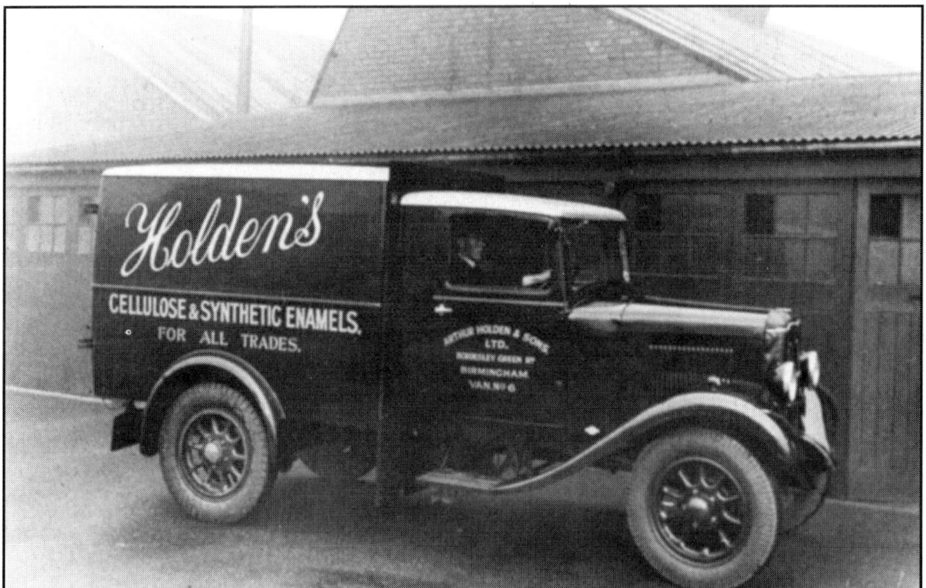

The vehicles of the Midland Counties Dairy were for many years a familiar sight on West Midlands roads. The company operated from several depots; this was necessary due to the limited range of horse drawn and electric powered vehicles. These vehicles were environmentally friendly as they didn't emit unpleasant fumes.

Two more Midland Counties vehicles. The horse drawn milk float was considered environmentally friendly by gardeners as they often collected the horses waste products for recycling on their rhubarb, roses etc.

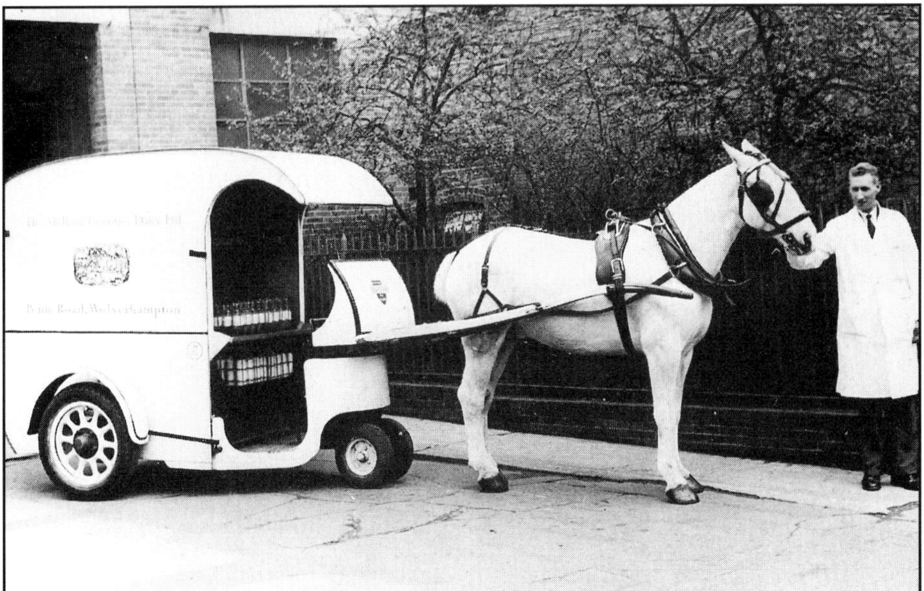

The original Morris Commercial one tonner, built at the Soho factory in 1924. This picture was taken during one of the many appearances of the vehicle in the London to Brighton historic commercial vehicle run. I drove this vehicle on several occasions.

This Morris Commercial was truly a wolf in sheeps clothing. The normal engine for this vehicle would have been grossly under powered Saurer diesel. This chassis was different as it was being used for proving the 5.1 litre engine that was eventually, in parallel with an uprated 5.7 litre version to become the mainstay of BMC five and seven ton vehicles for several years. The engines were developed and initially built at the Adderley Park plant in Birmingham, but production later transferred to Bathgate.

MCC LOGO OF 1940S

I make no excuse for my use of material associated with the former Morris Commercial cars. I remember seeing their vehicles from an early age, as they were then tested in the road in which we lived. The company no longer exists, it was in its day one of the principal manufacturers of light and medium commercial vehicles. One of its major customers was the Royal Mail. The demise of the company came with the Austin/Morris (BMC) rationalisation fiasco. The MCC plant at Adderley Park saw its work progressively transferred to other factories in the group. All work associated with heavy vehicles was moved to the white elephant plant at Bathgate. During the years that it took to complete rationalisation, BMC created a very costly network of material movement between plants. I started my working career as an apprentice at MCC in 1950. I was destined to remain with the company until 1971, finishing up as senior foreman in the transport department.

The ambulance is a Morris Commercial CVS11/30 with a coach built body. Pictured outside Saltley College which was adjacent to the companies East works. This part of the works had at one time been the home of SU carburettors.

The Oxford taxi cab was originally produced by Wolseley Motors, at their Drews Lane factory in Birmingham. The first production vehicle left the factory in February 1947. A prototype of the cab had been on the streets of London for several years prior to production commencing. During these years of proving, the vehicle clocked up in excess of 100,000 miles. By the time I joined Morris Commercial in 1950, rationalisation within the Nuffield organisation had resulted in taxi cab production being transferred to the MCC factory. By this time production was only running at a modest level, with each vehicle being hand assembled to the highest of engineering standards.

The style of house in the background would indicate that the picture of the Oxford cab. HON 324 was taken in the vicinity of the Drews Lane factory. The gentleman standing by the open passenger door is Lord Nuffield (William Morris). I got to know this cab well as it became attached to the MCC works transport fleet. During my time as car fleet mechanic I was responsible for its maintenance.

The Morris Commercial Royal Mail van is a 1934 model. The Royal Mail was very traditional with its specifications and many of the features of the 1934 model would have still been in the models in the 1950s.

The vehicle making doorstep deliveries for County Table Waters, is an Adderley Park built, Morris Commercial CF type.

Camping has been a popular pastime for many years. During the 1930s and 40s cars were a much rarer sight than nowadays. Most campers of that era relied on backpacking, public transport or cycling. The pictures on this page show the Bendelow family and their pre motoring rig. It is amazing how much equipment an experienced camper can pack in a small space. My family and I spent many happy times camping with the Bendelows in the 50s and 60s.

Motor coach travel has grown in popularity since its start in the days of the humble charabanc, which offered basic seating accommodation with little comfort or protection from the weather. By 1928 Smith's Imperial Coaches acquired a splendid Maudslay chassis which was fitted with a Buckingham coach built body and a sunshine roof. This vehicle remained in service throughout the 1930s.

The austere utility style vehicle was used by Boultons of Church Stretton to provide a rural bus service in South Shropshire.

Haymaking as it used to be. The mower is drawn by a pair of heavy horses, (gentle giants). A familiar sight in my youth. Probably at Brook House Farm, near the Drawbridge on the canal at Shirley.

The tractor is shown in its off road mode of yesteryear. Before the more recent advances of tyre technology, it was necessary to change tractor wheels for operating off road in fields.

A selection of local (Small Heath) adverts of the 1930s. (Piggotts was on the corner of Green Lane and Third Avenue).

Local garages/filling stations of yesteryear. Note the swinging gantry arms on the pumps, this was to allow vehicles to fill up without the pump hoses impeding the free passage of pedestrians.

A trunk road (A140) filling station and garage of yesteryear at Brockford Street, near Stowmarket. Note the old Ford van and large number of telegraph wires. Close scrutiny of the pumps shows that this garage was one of the few to sell diesel fuel at that time.

A small town oil distribution depot of yesteryear.

During my childhood, Birmingham was justly proud of the utility services provided to its citizens. These utilities existed due to the foresight of the city fathers in earlier years. Pictured below is a service van of the old Birmingham gas department. My late mother worked as an accounts clerk at the gas departments offices in Edmund street throughout the second world war.

A contractors vehicle removing solid treated sewage, possibly from the Minworth treatment works.

H Samuel press advert used from 1940 to 1945.

The H Samuel Coventry branch ready for Christmas trade in December 1932. The company at the time was under the control of the Edgar family. This picture is a good example of their trading strategy, which was to display as much as possible of their range of merchandise. This had the effect of making the shop inviting to customers. The policy obviously worked, as over the following few decades the company prospered and grew.

Morgan F4 family model. The prototype of this model was built during 1934/35. The model was in production from 1936/37 till 1950.

This was the first car I owned. A Morris 16 of 1935 vintage, bought for £40= in 1954.

Austin 7 two seater vintage 1934. Bought by me for £25= when I was in the RAF, 1956.

Birmingham trolley bus at Hay Mills terminus.

Birmingham bus of the 1930s. A 1929/30 AEC Regent with a 52 seat Brush body.

Birmingham tram. They offered a bumpy ride with little comfort for passengers. Particularly on the ones that had utility style slatted wooden seats.

Braggs confectionery shop that fronted their Alum Rock bakery, as it was in 1932. This shop front was to be the role model of Braggs shops for many years.

Barrage balloons were a familiar sight around the suburbs of Birmingham during the blitz of the early 1940s. I recall two balloons being sited on Hay Barn recreation ground and another in Digby park. On at least one occasion the latter broke free from its cable. The picture was taken 2.9.1940, "somewhere in England". The winch control vehicle is possibly a Fordson 79 of 1937/38 vintage. The trailer on the right is loaded with cylinders which would have contained hydrogen.

Heavy AA Battery in action during an air raid 14th October 1940. This battery was sited at Mount Kemel, near Dover. The guns of this battery would be similar to those sited in the park in Yardley, Birmingham and now known as the Oaklands Recreation Ground.

The Cyclo Gear Company was situated in Potters Hill, Aston and for many years produced a range of rugged multi speed gears that were popular with both touring and racing cyclists. My first cycle (a Sun, also made in Aston) was fitted with Cyclo gear and was bought from Whitworths in Corporation Street for the then (1948) huge sum of £19-10-6.

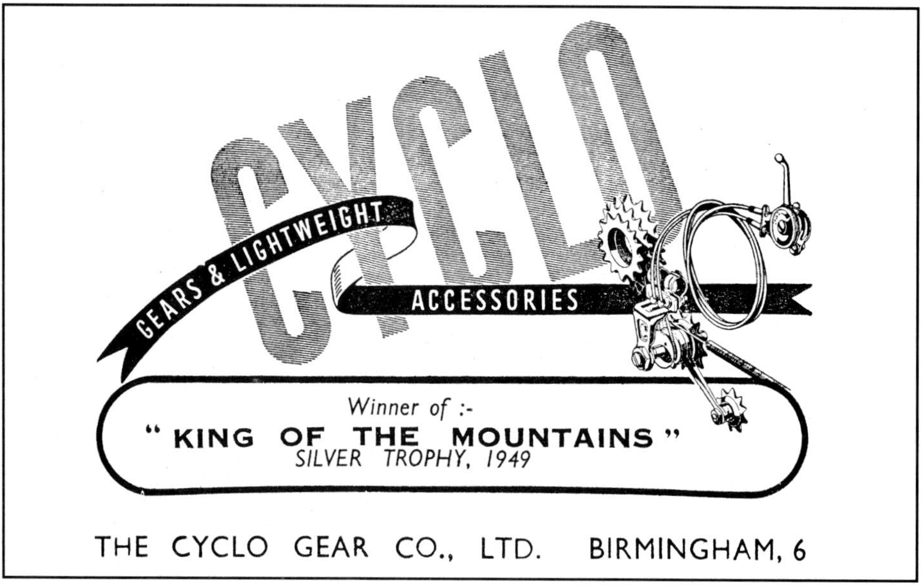

Barrow's Store provided the service and quality of yesteryear for those who could afford and were prepared to pay that little extra.

In the years immediately following the Second World War, great demands were made on builders. In addition to the urgent need to repair bomb damaged and destroyed premises there was a demand to have long neglected homes refurbished and modernised. The builders wagon pictured below was owned by George Harris and Sons and is shown in their yard at Kingston Road, just across the railway from St Andrews football ground. The Harris brothers were cousins of my mother and I have many childhood memories of hanging over their fence to watch the steam trains go by. They had a large house next to the yard, also a small sweet shop which got a lot of its custom from people going to the Kingston cinema, facing the Harris premises.

In the 1920s Walter Harris played for Birmingham City and was also one of the few private car owners of that time. He owned an Essex in which he drove my parents from their wedding in 1930.

Valor is a very old established Birmingham company with a large factory in Wood Lane, Erdington. The stoves illustrated depict the style of paraffin heaters of yesteryear. I remember this type of stove being the only form of heat in my grandfathers hardware shop. Also shown is a picture of the Valor assembly shop of the period.

"VALOR" Oil Heaters

USE "VALOR" WICKS ONLY.

Model **56 R** Model **525 R** Model **106 R**

The "VALOR" Works

My wife has vivid memories of horse drawn hearses, as her childhood home was in Yardley Road, facing the cemetery. She often speaks of seeing the black horses, adorned in black plumes as they drew the hearses.

Yardley Road, note the absence of traffic. Pre World War Two as the cemetery railings are still intact. The advert for White & Sons refers to the monumental sculptures who had a yard and works on Stockfield Road/ Yardley Road corner.

There was always great excitement when a travelling funfair arrived on the local recreation ground. This fine example of the type of showmans trailer of yesteryear was photographed by me in a garden at Richards Castle during 1995.

A touring caravan of the late 1940s. Photographed at Dymchurch in 1951. Statically sited at the time.

The Class of 1948, author with friends at Bordesely Green Tech. The master is the late Fred Markwell, a true gentleman whose lessons on the origins of surnames, stimulated my lifelong love of genealogy.

Birmingham City team of 1946/47 - from my boyhood autograph book

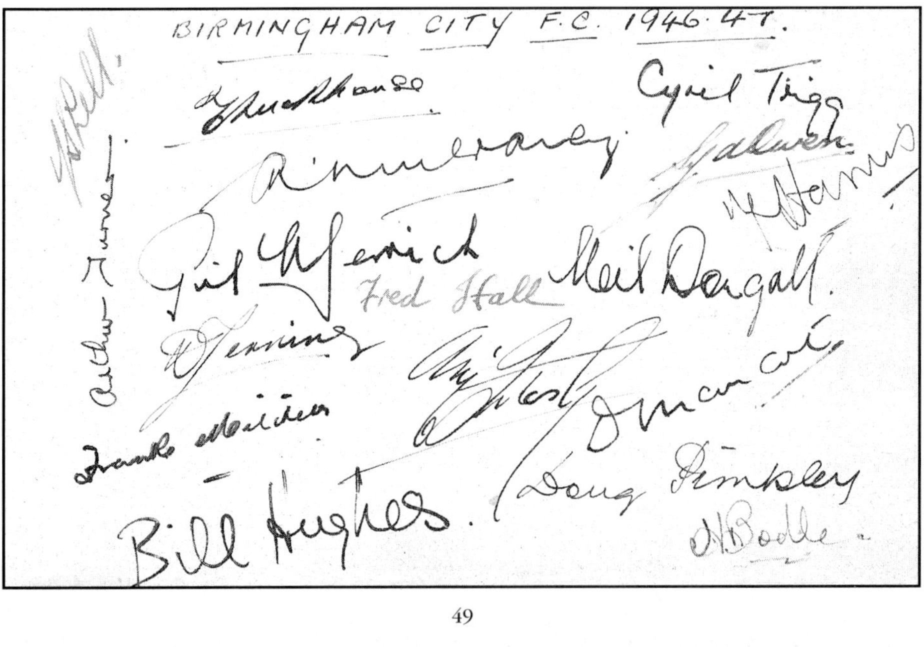

More adverts of the 1930s. Note the price of holidays in the Church Travellers' Club advert.

CHURCH TRAVELLERS' CLUB.

Chairman: LORD SAYE AND SELE.

OBER-AMMERGAU PASSION PLAY,

9 Days' Tour. OBER-AMMERGAU AND BRUGES OR HEYST, from **£10. 19s. 6d.** upwards.

Tours at varying rates including
COLOGNE, MUNICH, NUREMBERG, ROTHENBURG,
THE TYROL, THE BAVARIAN ALPS.

Special Illustrated Booklet.

Some specimen Tours.

INTERLAKEN (10 Days' Tour)	**£8**	**5**	**6**
MONTREUX—CLARENS (10 Days' Tour) ..	**£7**	**16**	**0**
LUCERNE (10 Days' Tour)	**£7**	**15**	**0**
NICE (10 Days' Tour)	**£11**	**10**	**0**
ROME (1 night Milan or Turin)	**£15**	**15**	**0**

GRAND CRUISE ROUND ITALY,

August 26th, September 9th and 23rd, October 7th, visiting Naples, Palermo, Taormina, Malta, Venice, etc. Inclusive costs, 19 days, 28 guineas and 35 guineas.

SUMMER HANDBOOK, 112 pages, Illustrated,
from SECRETARY,
(Dept. Y.), 3 ALBANY COURTYARD, PICCADILLY, LONDON, W.1.

Tel., VIC. 0211.

ROSAMUND BRETT.

Ladies' and Children's Expert Hairdresser,

70, HOBMOOR ROAD
(opposite St. Benedict's Road).

Permanent Waving, Tinting.
Manicuring, Shampooing, Beauty Culture.

SMITH'S IMPERIAL COACHES,

180, Stratford Road, Sparkbrook
(Garage)

ALL PARTIES LARGE & SMALL CATERED FOR.

Tel., VIC. 0668.

Where to obtain GOOD BREAD and CONFECTIONERY.....

HOUSE'S,

580, Green Lane, 155, Muntz St., Small Heath.
352, Stratford Road, Sparkhill.

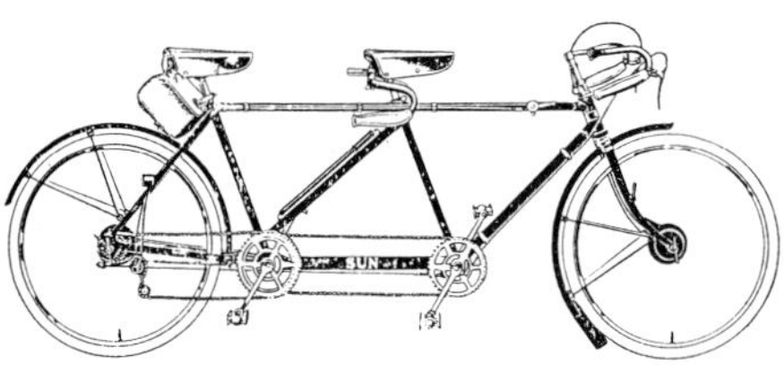

Visit England's Beauty Spots on a Bicycle
Supplied by

S. H. JONES

449 BORDESLEY GREEN (MEMBER A.E.U. SINCE 1903.)

Agent for B.S.A., Raleigh, Hercules Cycles and Motor Cycles. Cash or Easiest of Terms.
Renovations and Repairs. Accessoriesof every description. RADIOS supplied.
Accumulators Charged. Established 1908. Please Send Post Card for Catalogue.

A Works Outing of Yesteryear.

It took a great many 'wheels' on the coaches that transported this party of employees and friends from the High Street premises of the Birmingham Co-op to Dovedale in 1951.

This Excursion was on a wednesday afternoon, which was half day closing for the branch.

A final reminder of some of the trials & shortages of yesteryear. Many items remained in short supply and were the subject of rationing for several years after the finish of the Second World War (1939 to 1945).
The petrol coupons were for EON 850, a Austin 12. This was my fathers company car. It was eventually replaced with an Austin A40 Devon (complete with flying A) in 1951.

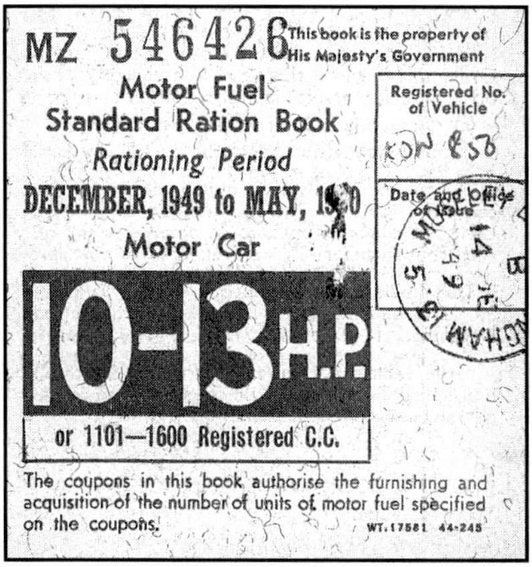

Many items of food remained on ration for a considerable time. In fact food rationing did not totally cease until shortly after I married in 1954.

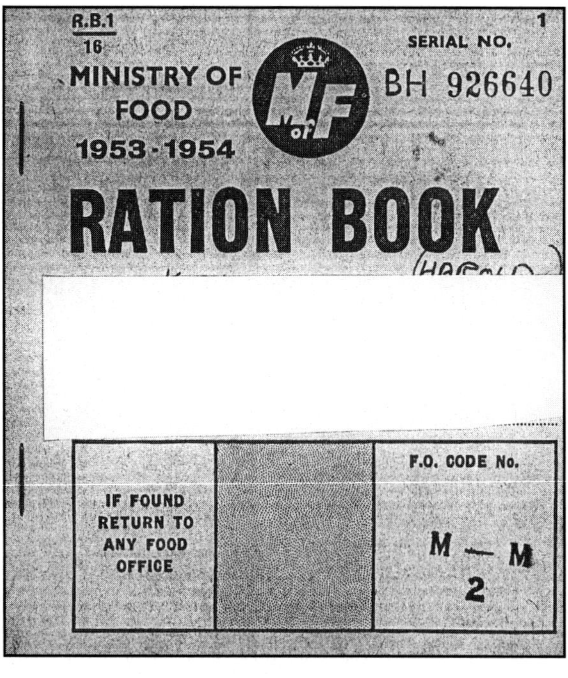

The author, still cycling in 1993

Pictured in front of his former home at Hollwood. Holding his Falcon cycle which was fitted with a Cyclo Benelux five speed gear and Williams chainset. All good sound and reliable British products.

EPILOGUE

When I read through my work, I was filled with a feeling of satisfaction (and relief). I felt that I had achieved about the right mix of 'Wheels & Things', which was my objective. It has been a pleasure to put this collection of nostalgia together. The task has been made more pleasurable by being able to work in my home environment, an environment that is in all ways superior to many that I have worked in over the years. During the time I have been writing this book I have had to concede that advancing years, a heart condition and not least a home which is 1000 feet up the side of a hill that it would be sensible to finally hang up my cycle clips; so I have.

Eric W Russell,
Clee Hill,
Shropshire.

August 1996

AND FINALLY

During the 1940s there was a cartoon type character on the scene who went by the name of 'MR. CHAD', commonly referred to as 'CHAD'. He appeared on posters, on walls, in magazines and newspapers. He even at times appeared on toilet walls with some unprintable captions. He was normally shown highlighting shortages and problems of the time. This was achieved by having a caption that asked a question. The way in which he was presented often helped to make light of the problems he portrayed.

ACKNOWLEDGMENTS

I am indebted to the companies and private individuals whose cooperation has made it possible for me to put this collection of pictures together. In response to my pleas, (requests) for assistance, most of those whom I approached proved positive with their responses. A few failed to respond either way and a small number put the price of cooperation above my limited resources. I reserve my thoughts on the large multi national organisations who reacted in this way.

I have attempted to make specific mention of all who have provided material for me to use. If I have missed anyone from the following list, I offer them my humble apologies.

Special thanks are due to Peter Bartlett of Ludlow for his work in making photographic copies of original material loaned to me. Much of which was in very poor condition when received from the owners. Also to my son in law, Geoffrey Heath, who armed only with a very rough pencil sketch I gave him, did the graphics for the cover of this book. Finally a special mention of my dear wife, Mary, who has shown great patience and understanding during the time I have devoted to this project.

Original material loaned by:
Norman Painting, the Post Office, Harold Bendelow, Mrs Enid Keech, Smiths Imperial Coaches, Boultons of Shropshire, Esso UK plc., Ford Motor Company, Severn Trent Water, British Gas, Braggs Bakers, Signet Group plc., Birmingham Post & Mail, Morgan Motor Company, West Midland Travel, The Airship Heritage Trust, Scotts Cycles, Imperial War Museum, Birds Eye Wall's, Francis Benton, Lucas Ltd., The Automobile Association, The Royal Automobile Club, Vale Onslow, Udall Family, Kraft-Jacobs-Suchard Ltd., R H Collier, Alf Jenkins (Titterstone Clee Hill), Avonmore Dairies Ltd., Small Heath Amateur Gardeners Club, Lee Longland, Cadbury Ltd., SP Tyres UK Ltd., Holden Surface Coatings Ltd., Vic Harris, Valor Heating, Pugh & Sons (undertakers), Mrs Vera Heard.